AI SUPERPOWERS COMPETITION

The Impact of AI on Economy and Society

Patrick Odega

ISBN: 9798394815348

Dedication

This book is dedicated to the researchers, engineers, and innovators who are driving progress in the field of artificial intelligence. Their work has the potential to transform industries and improve lives around the world. May their efforts lead to the development of AI that is ethical, responsible, and serves the greater good.

Table of Contents

Introduction:

Artificial intelligence, or AI, is one of the most transformative technologies of our time, with the potential to revolutionize many aspects of our lives, from healthcare and transportation to finance and education. AI involves the development of algorithms and systems that can perform tasks that typically require human intelligence, such as recognizing speech, understanding natural language, and making decisions.

The global race for AI dominance is intensifying, with the United States and China at the forefront of this competition. Silicon Valley, the hub of innovation in the US, has a long history of leadership in AI research and development, while China has made rapid progress in recent years and is now investing heavily in AI as a strategic priority. The potential implications of this AI race are vast, with the potential to shape the future of the global economy, politics, and society.

In this book, we will explore the rise of AI, the unique advantages and challenges facing China and Silicon Valley, the competition and collaboration between them, and the implications of AI for society. Through this exploration, we hope to gain a deeper understanding of the opportunities and risks of AI, and the role of China and Silicon Valley in shaping the future of this transformative technology.

Chapter 1:

China's AI Awakening

China has set a goal to become the world leader in AI by 2030, and is investing heavily in research and development, talent acquisition, and infrastructure to achieve this goal. The Chinese government has launched several national plans to support the development of AI, including the New Generation Artificial Intelligence Development Plan and the Made in China 2025 initiative. These plans aim to promote research and development, talent acquisition, and innovation in AI, as well as to build up infrastructure such as data centers and 5G networks to support AI applications.

China's progress in AI has been impressive, with significant breakthroughs in areas such as natural language processing, computer vision, and robotics. Chinese tech giants such as Alibaba, Baidu, and Tencent have made significant investments in AI, and Chinese startups are also emerging as leaders in this field.

The Advantages and Challenges of China's AI Development

China has several advantages in the field of AI, including a large and growing population, abundant data, and government support for innovation. China's massive population provides a large pool of data that can be used to train AI algorithms, and its government is investing heavily in AI to drive economic growth and innovation.

However, China also faces several challenges in its AI development, including issues related to data privacy and intellectual property. China's regulatory environment for data privacy is not as robust as that in the US and Europe, which could hinder the development of AI applications that rely on personal data. Intellectual property theft is also a concern, as some Chinese companies have been accused of stealing technology and trade secrets from foreign competitors.

The Impact of AI on China's Economy and Society

The impact of AI on China's economy and society is already significant, with AI transforming industries such as healthcare, transportation, and finance. For example, AI is being used to improve medical diagnoses, enhance transportation safety, and optimize financial investments.

AI is also expected to drive significant economic growth in China, with some estimates suggesting that AI could add trillions of dollars to the Chinese economy by 2030. However, the impact of AI on employment is a concern, as some jobs may be replaced by automation. The Chinese government has launched several initiatives to address this issue, including retraining programs for workers and policies to encourage the development of new industries.

China's AI awakening represents a significant challenge and opportunity for the global AI landscape. The country's massive investment in AI and data infrastructure is likely to accelerate the pace of AI development and innovation, but it also raises important questions about data privacy, intellectual property, and the ethical implications of AI.

Chapter 2:

Silicon Valley's AI Dominance

Silicon Valley has a long history of leadership in the field of AI. Some of the earliest work in AI was conducted in the region in the 1950s and 1960s at institutions such as Stanford University and the Massachusetts Institute of Technology (MIT). In the 1980s and 1990s, Silicon Valley startups such as Apple and Hewlett-Packard played a key role in the development of AI technologies such as expert systems and natural language processing.

In recent years, Silicon Valley has emerged as a hub of innovation in AI, with companies such as Google, Facebook, and Microsoft investing heavily in AI research and development. Silicon Valley has also attracted top AI talent from around the world, and has become a magnet for startups and venture capital investment in the field of AI.

The Culture and Talent that Drive Innovation in Silicon Valley

Silicon Valley's culture of innovation and risk-taking has been a key driver of its leadership in AI. The region's startup culture, with its emphasis on rapid prototyping and iteration, has allowed AI researchers and developers to experiment with new ideas and technologies in a fast-paced environment.

Silicon Valley's talent pool is also a major advantage in the AI race. The region attracts top AI talent from around the world, and its universities and research institutions are home to some of the most innovative AI researchers and engineers.

The Impact of AI on Silicon Valley's Economy and Society

AI is transforming industries such as tech, healthcare, and finance in Silicon Valley. For example, AI is being used to improve search algorithms, personalize marketing, and enhance cybersecurity.

AI is also driving significant economic growth in Silicon Valley, with some estimates suggesting that AI could add trillions of dollars to the US economy by 2030. However, the impact of AI on employment is a concern, as some jobs may be replaced by automation. Silicon Valley has launched several initiatives to address this issue, including retraining programs for workers and policies to encourage the development of new industries.

Overall, Silicon Valley's dominance in the field of AI represents a significant opportunity and challenge for the global AI landscape. The region's culture of innovation and talent pool are likely to continue to drive AI development and innovation, but it also raises important questions about the ethical implications of AI and the need for responsible AI development and governance.

The Future of AI Development and Governance in Silicon Valley

As Silicon Valley maintains its position as a global AI leader, the future of AI development and governance becomes increasingly important. The region faces both exciting opportunities and critical challenges as AI continues to evolve and shape society.

One key aspect of the future of AI development is the pursuit of cutting-edge technologies and breakthroughs. Silicon

Valley's culture of innovation and risk-taking drives the relentless pursuit of advancements in AI. Researchers and engineers are continually pushing the boundaries of what AI can achieve, exploring new algorithms, architectures, and applications. The development of AI models capable of higher levels of understanding, reasoning, and adaptability remains a primary focus.

In parallel, the responsible and ethical development of AI is a critical consideration. As AI becomes more integrated into various aspects of daily life, ensuring transparency, fairness, and accountability is paramount. Silicon Valley is increasingly aware of the ethical implications of AI and the need to address biases, privacy concerns, and the potential impact on jobs and society. Efforts are being made to establish ethical frameworks, guidelines, and regulations that guide the development and deployment of AI technologies.

Collaboration and partnerships play a crucial role in shaping the future of AI in Silicon Valley. The region thrives on an ecosystem of collaboration, where companies, academic institutions, and startups work together to tackle complex challenges. Collaborative efforts foster knowledge exchange, interdisciplinary research, and the pooling of resources, leading to more innovative and impactful AI solutions.

The collaboration extends beyond regional boundaries. Silicon Valley actively engages in global collaborations, partnering with international organizations, researchers, and governments. By sharing expertise and insights, collaborating on research projects, and establishing international standards, Silicon Valley can contribute to a more inclusive and globally beneficial AI ecosystem.

The Future Impact of AI on Silicon Valley's Economy and Society

The impact of AI on Silicon Valley's economy and society is expected to continue evolving. AI-driven automation and efficiency improvements have the potential to reshape industries and work dynamics. While some jobs may be displaced by automation, new opportunities may arise as AI creates new industries and transforms existing ones. Silicon Valley is poised to adapt and capitalize on these changes, driving economic growth through the development of AI-powered products and services.

In society, the integration of AI may bring about both benefits and challenges. AI technologies have the potential to enhance healthcare outcomes, revolutionize transportation systems, optimize energy consumption, and tackle pressing global issues. However, concerns such as privacy, algorithmic bias, and the impact on human decision-making need to be addressed. Silicon Valley must take a proactive role in ensuring that AI is developed and deployed in ways that align with societal values and serve the greater good.

The future of AI in Silicon Valley encompasses ongoing advancements in technology, the responsible and ethical development of AI, collaboration and partnerships, and the impact on the economy and society. As Silicon Valley continues to drive AI innovation, it must navigate the complex challenges and seize the opportunities to shape a future where AI benefits all of humanity.

The Ethics of AI Development and Use in Silicon Valley

As Silicon Valley maintains its AI dominance, the ethical implications of AI development and use have become central to discussions and actions within the region. The responsible and ethical development of AI technologies is crucial to ensure that the potential benefits are maximized while minimizing potential risks and unintended consequences.

Ethical considerations in AI development encompass various aspects, including fairness, transparency, accountability, and privacy. Silicon Valley acknowledges the importance of addressing biases that can be embedded in AI algorithms and data, ensuring that AI systems are fair and do not perpetuate discriminatory practices. Efforts are underway to improve transparency in AI decision-making, enabling users to understand how AI systems arrive at their conclusions. Moreover, accountability mechanisms are being explored to ensure that the responsible parties can be held responsible for the actions and outcomes of AI systems.

Privacy is another key ethical concern in AI development. Silicon Valley is working on developing privacy-preserving AI technologies and implementing robust data protection measures to safeguard individuals' privacy rights. Striking a balance between utilizing data for AI advancements and protecting individuals' privacy is an ongoing challenge that requires continual attention and innovation.

Additionally, AI safety and security are important ethical considerations. As AI systems become more autonomous and capable of making critical decisions, ensuring their safety and security is crucial. Silicon Valley invests in research and development efforts to mitigate risks and prevent malicious use of AI technologies. Collaboration among industry, academia, and policymakers is essential to establish standards, guidelines, and regulations to govern the development and deployment of safe and secure AI systems.

Education and awareness about the ethical implications of AI are also important aspects of Silicon Valley's approach. Efforts are being made to educate AI developers, researchers, and users about the potential ethical challenges and responsibilities associated with AI technologies. Encouraging interdisciplinary collaboration and engaging in public discourse around AI ethics are essential for ensuring that the development and use of AI align with societal values and promote the common good.

The Role of International Collaboration in the Global AI Landscape

Silicon Valley recognizes the need for international collaboration in the field of AI. The global nature of AI challenges and the potential impact of AI technologies on a global scale require collective efforts and cooperation among countries, organizations, and researchers.

International collaboration in AI research and development fosters knowledge sharing, facilitates the exchange of best practices, and promotes the adoption of ethical and responsible AI standards globally. Collaborative initiatives can help address challenges such as data sharing, interoperability, and standardization, allowing for more effective and beneficial AI applications across borders.

Furthermore, international collaboration contributes to a more inclusive and diverse AI ecosystem. By involving a wide range of perspectives and experiences, AI technologies can be developed in a manner that considers the needs and values of different societies and cultures. This collaboration can also help bridge the AI divide, ensuring that countries and regions with fewer resources or expertise in AI can benefit from advancements and participate in the global AI landscape.

To foster international collaboration, Silicon Valley actively engages with global partners, participates in international conferences and forums, and supports initiatives that promote collaboration in AI research and development. By working together, sharing knowledge, and aligning efforts, the global community can collectively address the challenges and seize the opportunities presented by AI technologies.

As Silicon Valley continues to lead in AI development, the region recognizes the importance of ethical considerations, international collaboration, and the global impact of AI technologies. By prioritizing responsible AI development, addressing ethical challenges, promoting international collaboration, and engaging in public discourse, Silicon Valley can contribute to the advancement of AI technologies that benefit humanity while ensuring that AI is developed and used in ways that align with societal values and promote the greater good.

Chapter 3:

The AI Competition and Collaboration between China and Silicon Valley

The race for AI supremacy between China and Silicon Valley has garnered significant attention due to the technological advancements and economic potential associated with AI. Both China and Silicon Valley have emerged as major players in AI research, development, and innovation, each with its unique strengths and weaknesses.

China's Rise in AI Development:

China has experienced a remarkable rise in AI development in recent years, driven by various factors. One of China's strengths lies in its large population, which provides a vast pool of data for training AI models. This abundance of data allows Chinese companies and researchers to develop AI systems that are tailored to the Chinese market and its specific needs.

Additionally, the Chinese government has made AI a national priority, investing heavily in research and development, infrastructure, and talent cultivation. Government initiatives, such as the "Made in China 2025" plan and the "New Generation Artificial Intelligence Development Plan," have provided financial support and incentives for AI development, attracting both domestic and international investments.

China's advantage also stems from its strong presence in the technology and manufacturing sectors. Chinese tech giants, including Baidu, Alibaba, and Tencent, have made significant investments in AI research and development, driving innovation in areas such as natural language processing, computer vision, and autonomous vehicles. Moreover, China's manufacturing capabilities enable rapid deployment and scalability of AI technologies in various industries.

Silicon Valley's Dominance in AI:

Silicon Valley, long recognized as a global hub for technology and innovation, has a deep-rooted leadership position in AI. The region benefits from a rich ecosystem that fosters entrepreneurial spirit, risk-taking, and a culture of innovation. This environment attracts top AI talent and encourages collaboration and cross-pollination of ideas among researchers, startups, and established tech companies.

Silicon Valley's strength in AI development lies in its expertise in advanced algorithms, machine learning, and deep learning techniques. It has a history of pioneering breakthroughs in AI research, dating back to the early days of the field. Renowned universities, such as Stanford and UC Berkeley, and research institutions, such as OpenAI and the Allen Institute for Artificial Intelligence, contribute to the region's intellectual capital and drive AI advancements.

Moreover, Silicon Valley has a strong venture capital ecosystem, allowing startups and researchers to access funding for AI projects.

This financial support enables experimentation, innovation, and the translation of AI research into practical applications. The region is home to leading AI-focused companies, such as Google, Facebook, and Tesla, that continually push the boundaries of AI capabilities and drive industry trends.

The Competition for AI Supremacy:

The competition for AI supremacy between China and Silicon Valley extends beyond technological advancements and economic gains. It also carries geopolitical implications, as AI is considered a key driver of future economic growth, national security, and global influence.

China's government-led approach to AI development has allowed it to make rapid progress and deploy AI technologies at a large scale within its domestic market. China's strengths in areas such as facial recognition, smart cities, and e-commerce have given it a competitive edge in certain AI applications. The country's focus on AI innovation and strategic investments in emerging technologies like 5G and quantum computing position it as a formidable competitor.

On the other hand, Silicon Valley's long-standing leadership in AI research, talent pool, and entrepreneurial ecosystem gives it an advantage in developing cutting-edge AI algorithms and pushing the frontiers of AI capabilities. The region's focus on high-risk, high-reward innovation and its vibrant startup culture foster continuous AI advancements and disruptive breakthroughs.

Potential Consequences of the AI Competition:

The AI competition between China and Silicon Valley has potential consequences at multiple levels. Economically, the race for AI supremacy can drive economic growth, job creation, and industry transformation in both regions. The development and deployment of AI technologies can lead to increased productivity, efficiency gains, and the emergence of new AI-driven industries. However, the competition also raises concerns about job displacement and growing inequalities, as certain sectors and occupations may be more vulnerable to automation.

Geopolitically, the AI competition between China and Silicon Valley has implications for global power dynamics and influence. AI is seen as a strategic technology that can provide countries with a competitive advantage in various sectors, including defense, healthcare, and transportation. The winner of the AI race could potentially shape the rules and norms surrounding AI development, governance, and international cooperation.

Moreover, the competition raises ethical considerations, such as data privacy, algorithmic bias, and the impact of AI on human rights and civil liberties. Both China and Silicon Valley must address these concerns to ensure that AI technologies are developed and deployed in a manner that respects individuals' rights and promotes the well-being of society.

Possibilities for Collaboration and Cooperation:

Despite the competitive nature of the AI race, there are also opportunities for collaboration and cooperation between China and Silicon Valley. Both sides have unique strengths and expertise that, if combined, can drive innovation and address common challenges.

Collaboration in AI research can foster knowledge exchange and accelerate advancements. Sharing research findings, data sets, and best practices can enhance the development of AI technologies, particularly in areas where collaboration can lead to breakthroughs. Joint research projects and partnerships between Chinese and Silicon Valley-based institutions can leverage complementary expertise and resources.

International standards and guidelines are another area where collaboration is crucial. Establishing common ethical frameworks, privacy regulations, and safety standards can promote responsible AI development and ensure the interoperability and compatibility of AI systems across borders. Collaboration in policy discussions and sharing regulatory experiences can help shape a global AI governance framework that balances innovation, ethics, and societal considerations.

Furthermore, collaboration can extend beyond research and policy to include industry partnerships and investment opportunities. Silicon Valley companies can explore partnerships with Chinese counterparts to access the vast Chinese market, while Chinese companies can benefit from Silicon Valley's expertise and market insights. Joint ventures and cross-border investments can drive innovation, technology transfer, and commercialization of AI applications.

Ultimately, collaboration and cooperation between China and Silicon Valley in AI can lead to mutually beneficial outcomes. By leveraging their respective strengths, sharing knowledge, and working together, the two sides can address common challenges, promote responsible AI development, and unlock the full potential of AI for the benefit of humanity.

In conclusion, the AI competition between China and Silicon Valley is characterized by their unique strengths, research capabilities, and economic advantages. The competition has implications for economic growth, geopolitics, and societal implications. However, there are also possibilities for collaboration and cooperation, where both sides can leverage their strengths, share knowledge, and address common challenges. By finding areas of synergy and working together, China and Silicon Valley can contribute to the advancement of AI technologies while ensuring ethical considerations and promoting the greater good.

Chapter 4:

The Future of AI and its Implications for Society

Artificial Intelligence (AI) has the potential to transform society in profound ways, revolutionizing industries, enhancing productivity, and shaping our daily lives. However, as AI continues to advance, it brings with it a range of implications that need to be carefully considered to ensure responsible and beneficial development and deployment.

The Transformative Power of AI:

AI technologies have the ability to augment human capabilities and automate tasks, leading to increased efficiency and productivity across various sectors. From healthcare to transportation, finance to education, AI has the potential to revolutionize industries, streamline processes, and improve outcomes.

AI-powered systems can analyze vast amounts of data and extract valuable insights, enabling more informed decision-making. This can lead to advancements in fields such as personalized medicine, precision agriculture, and smart cities, where AI can optimize resource allocation, improve operational efficiency, and enhance quality of life.

Moreover, AI has the potential to address complex societal challenges. It can contribute to the development of solutions for climate change, poverty, and inequality by enabling better prediction, modeling, and decision-making. AI-driven technologies such as natural language processing and computer vision can also enhance accessibility and inclusivity by providing tools for people with disabilities.

The Ethical and Social Challenges:

While AI offers tremendous potential benefits, it also presents significant ethical and social challenges that need to be addressed. One of the key concerns is algorithmic bias, where AI systems can inadvertently perpetuate or amplify existing biases in data and decision-making. This can lead to discriminatory outcomes in areas such as hiring, lending, and criminal justice. Addressing bias and ensuring fairness in AI algorithms is essential to prevent harm and promote equal opportunities.

Privacy is another critical concern. AI systems rely on vast amounts of personal data, raising questions about data collection, consent, and protection. Balancing the benefits of AI with individuals' privacy rights requires robust data governance frameworks, transparency, and responsible data practices.

The impact of AI on employment is also a significant societal concern. While AI can automate certain tasks and lead to job displacement, it also has the potential to create new job opportunities. Preparing the workforce for the changing job landscape and ensuring equitable access to AI-driven employment opportunities are important considerations.

AI also raises questions about accountability and responsibility. As AI systems become more autonomous and make decisions that impact human lives, understanding the decision-making process and ensuring accountability become crucial. Clear guidelines, regulations, and oversight mechanisms are necessary to establish responsibility and address potential risks.

The Need for International Cooperation and Regulation:

Given the global nature of AI and its potential impact on society, international cooperation and regulation are vital. Collaborative efforts can ensure that AI is developed and deployed in ways that are aligned with shared values, ethical principles, and human rights.

International cooperation can foster knowledge sharing, best practices, and standards development. It can facilitate the exchange of expertise and experiences, enabling countries and organizations to learn from one another and collectively address challenges such as bias, privacy, and accountability.

Regulation is essential to provide a framework for responsible AI development. Governments and regulatory bodies play a critical role in establishing guidelines, standards, and oversight mechanisms to ensure the ethical and safe use of AI technologies. This includes considerations such as transparency, explainability, fairness, and accountability. Collaboration between governments, industry, and civil society can lead to the creation of comprehensive and effective regulatory frameworks.

In addition to cooperation and regulation, ongoing public engagement and dialogue are necessary to ensure that AI technologies are developed and deployed in a manner that aligns with societal values and aspirations. This includes involving diverse perspectives, soliciting public input, and addressing concerns and expectations regarding the impact of AI on individuals and communities.

The future of AI holds great promise and potential for transforming society. However, addressing the ethical, social, and regulatory challenges associated with AI development and deployment is crucial. It requires a holistic approach that involves stakeholders from governments, industry, academia, and civil society.

To ensure the responsible development of AI, transparency and explainability are essential.

AI systems should be designed in a way that allows users to understand the logic and decision-making processes behind their actions. This promotes trust and helps mitigate concerns about AI's black box nature.

Fairness and accountability should be core principles in AI development. Algorithms must be designed to avoid bias and discrimination, and mechanisms should be in place to address any unintended consequences. Additionally, establishing clear lines of responsibility and accountability for AI systems is necessary to address issues that may arise.

Privacy protection is paramount in the age of AI. Robust data governance frameworks should be implemented, ensuring that individuals have control over their personal information and that data is handled securely. Strict regulations and standards should be in place to safeguard privacy rights and prevent misuse of personal data.

Education and workforce development are also crucial aspects of the future of AI. As the job landscape evolves, individuals need to be equipped with the necessary skills to thrive in an AI-driven world. Governments, educational institutions, and industry should collaborate to provide training programs and reskilling opportunities to ensure a smooth transition and reduce the risk of job displacement.

International cooperation and collaboration are imperative to address the global challenges and implications of AI. Multilateral efforts can establish common ethical guidelines, technical standards, and norms for responsible AI development. This includes sharing research, best practices, and regulatory experiences to foster a collective understanding of the risks and benefits associated with AI.

Furthermore, collaboration between countries can help bridge the AI divide and ensure that the benefits of AI are accessible to all. It is crucial to address the global digital divide, as unequal access to AI technologies can exacerbate existing inequalities and hinder progress. Developing countries should be supported through capacity building initiatives, technology transfer, and investment to ensure inclusivity and a more equitable distribution of AI benefits.

The future of AI holds immense potential to positively impact society. However, it is essential to navigate its development and deployment responsibly. This requires addressing ethical considerations, ensuring transparency and fairness, protecting privacy rights, investing in education and workforce development, and fostering international cooperation and collaboration. By doing so, we can harness the transformative power of AI while safeguarding the well-being and values of individuals and communities worldwide.

Conclusion:

The future of AI holds great promise and potential for transformative advancements across various sectors. Both China and Silicon Valley play crucial roles in shaping the trajectory of AI development and deployment, each bringing their unique strengths and perspectives to the table.

China's Influence on the Future of AI:

China has emerged as a major player in the AI landscape, driven by its ambitious AI development plans, substantial investments, and a growing pool of AI talent. China's government-led approach to AI development has allowed it to make significant progress in areas such as facial recognition, natural language processing, and autonomous vehicles.

China's vast market size and data availability provide a fertile ground for AI applications and innovations. Chinese companies, backed by government support and investment, are actively deploying AI technologies across various sectors, including healthcare, finance, transportation, and e-commerce. This widespread adoption of AI in China positions the country as a leading testbed for real-world AI applications and provides valuable insights for future developments.

Furthermore, China's emphasis on AI research and education has resulted in the growth of AI-focused universities, research institutions, and tech companies. The collaborations between academia, industry, and government in China foster an ecosystem that promotes innovation, talent development, and knowledge sharing.

Silicon Valley's Role in Shaping AI's Future:

Silicon Valley, renowned for its innovation and entrepreneurial spirit, has been at the forefront of AI research and development for decades. The region's robust ecosystem of tech companies, startups, venture capital firms, and leading research institutions has been instrumental in pushing the boundaries of AI capabilities.

Silicon Valley's strengths lie in its expertise in advanced algorithms, machine learning techniques, and deep learning architectures. The region has been a breeding ground for AI breakthroughs, including significant advancements in areas such as computer vision, natural language processing, and reinforcement learning. The research and development conducted in Silicon Valley continue to drive AI advancements globally.

Moreover, Silicon Valley's entrepreneurial culture and risk-taking mindset have contributed to the rapid commercialization of AI technologies. Startups in the region have been instrumental in developing innovative AI applications and driving industry disruptions. The availability of venture capital funding and access to a vibrant ecosystem of mentors, advisors, and potential collaborators further fuel the growth of AI startups in Silicon Valley.

Collaboration between China and Silicon Valley:

The collaboration and exchange of ideas between China and Silicon Valley are crucial in shaping the future of AI. Both regions have unique strengths and perspectives that can complement each other, leading to synergistic advancements and addressing global challenges.

China's large-scale deployment of AI technologies and access to vast amounts of data can provide valuable insights for researchers in Silicon Valley. The collaboration can foster knowledge exchange, research partnerships, and joint projects that drive innovation and advancements in AI algorithms, models, and applications.

Silicon Valley's expertise in cutting-edge AI research, algorithm development, and commercialization can offer valuable guidance to China's AI ecosystem. Collaboration can focus on areas such as data privacy, algorithmic fairness, and ethical considerations, where sharing best practices and experiences can contribute to responsible AI development.

Furthermore, collaboration can extend to industry partnerships, investment opportunities, and technology transfer. Silicon Valley companies can explore collaborations with Chinese counterparts to access the Chinese market, while Chinese companies can benefit from Silicon Valley's expertise and global market reach. Joint ventures, investments, and cross-border collaborations can accelerate AI development, commercialization, and the widespread adoption of AI technologies.

In conclusion, the outlook for AI is highly promising, with the potential to revolutionize industries and positively impact society. China and Silicon Valley both play vital roles in shaping AI's future, leveraging their respective strengths and expertise. Collaboration and exchange between these two influential regions can lead to synergistic advancements, address global challenges, and ensure responsible and beneficial